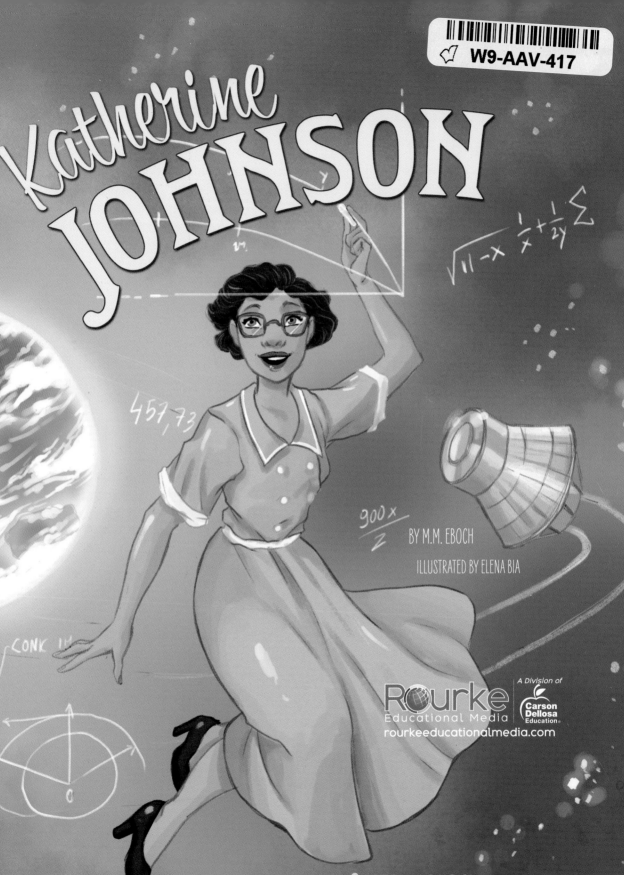

Katherine JOHNSON

BY M.M. EBOCH

ILLUSTRATED BY ELENA BIA

Rourke
Educational Media

A Division of
Carson
Dellosa
Education

rourkeeducationalmedia.com

WOMEN IN SCIENCE & TECHNOLOGY

Before Reading: *Building Background Knowledge and Vocabulary*

Building background knowledge can help children process new information and build upon what they already know. Before reading a book, it is important to tap into what children already know about the topic. This will help them develop their vocabulary and increase their reading comprehension.

Questions and Activities to Build Background Knowledge:

1. Look at the front cover of the book and read the title. What do you think this book will be about?
2. What do you already know about this topic?
3. Take a book walk and skim the pages. Look at the table of contents, photographs, captions, and bold words. Did these text features give you any information or predictions about what you will read in this book?

Vocabulary: *Vocabulary Is Key to Reading Comprehension*

Use the following directions to prompt a conversation about each word.

- Read the vocabulary words.
- What comes to mind when you see each word?
- What do you think each word means?

> **Vocabulary Words:**
> - *analyzed*
> - *geometry*
> - *integration*
> - *orbit*
> - *physics*
> - *research*
> - *segregation*
> - *servant*

During Reading: *Reading for Meaning and Understanding*

To achieve deep comprehension of a book, children are encouraged to use close reading strategies. During reading, it is important to have children stop and make connections. These connections result in deeper analysis and understanding of a book.

Close Reading a Text

During reading, have children stop and talk about the following:

- Any confusing parts
- Any unknown words
- Text to text, text to self, text to world connections
- The main idea in each chapter or heading

Encourage children to use context clues to determine the meaning of any unknown words. These strategies will help children learn to analyze the text more thoroughly as they read.

When you are finished reading this book, turn to the next-to-last page for **Text-Dependent Questions** and an **Extension Activity**.

TABLE OF CONTENTS

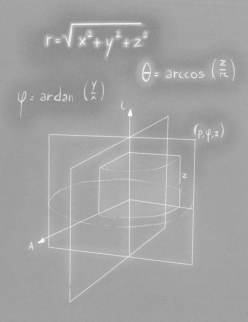

KATHERINE COUNTS

Young Katherine loved to count. She counted the steps to the road. She counted the steps up to church. She counted the forks and plates when she washed dishes.

Smart and curious, she skipped several grades in school. By age ten, she was ready for high school!

But her town did not have a high school for African Americans. They were not allowed to go to school with white children.

Her father took the family to another city. He said, "You are as good as anybody in this town." He left them there and went back home to farm. Their mother worked as a **servant**. The children stayed in school.

MATH FOR SPACE

At 15, Katherine went to college. She learned all about math. One teacher made a class just for her. It covered **geometry** in outer space.

Katherine finished college with high honors. She had degrees in mathematics and French. Few women had math degrees. Few African Americans had jobs in math.

$$r = \sqrt{x^2 + y^2 + z^2}$$

$$\theta = \arccos\left(\frac{z}{r}\right)$$

$$\varphi = \arctan\left(\frac{y}{x}\right)$$

(ρ, φ, z)

(r, θ, ρ)

West Virginia State
Katherine went to West Virginia State University. Only African Americans went there. Many West Virginia schools were separated by race. In 1939, the state began mixing some schools. This was called **integration**.

Katherine taught at a school for African Americans. Then, West Virginia University started taking African American students. Katherine went back to college and studied math again. She was the only African-American woman there.

Katherine married James Goble. They had three daughters. Then James got sick. Katherine went back to teaching to earn money.

A government group wanted women math experts. Katherine and James moved so she might get a job. They did not hire her.

Later, Katherine tried again. They hired her! She worked with a large group of women. They used rulers and adding machines to solve hard math problems. Once again, Katherine was counting.

In 1953, Katherine began work at a new lab. She **analyzed** flight tests. For one job, she investigated a plane crash.

Katherine wanted to know how and why things worked, or why they didn't. Only men were supposed to go to meetings. Katherine went anyway. She met more people and learned new things. She became an important team member.

"I'LL DO IT BACKWARDS"

Katherine's husband died in 1956. The next year the Soviet Union sent the first satellite into space. This pushed America to grow its space program. They needed Katherine's math more than ever.

NASA wanted to know how spacecraft could go around Earth. This used a lot of complex math! Katherine said, "Let me do it. You tell me when you want it and where you want it to land, and I'll do it backwards and tell you when to take off."

NACA to NASA
Katherine worked for NACA (National Advisory Committee for Aeronautics). There, African-American women used separate bathrooms. They ate in separate dining rooms. That changed when NACA became NASA (National Aeronautics and Space Administration). **Segregation** stopped.

In 1959, Katherine married James Johnson. She became Katherine Johnson.

Her career was only beginning. Machines now did the math for space flights. But computers made mistakes. Astronaut John Glenn didn't trust them. For his Friendship 7 mission, he asked Katherine to check the numbers. She used a mechanical adding machine. It worked with wheels and levers. When she said the numbers were right, Glenn was happy to go. He became the first American to **orbit** Earth.

Katherine worked at NASA for 33 years. She wrote many **research** reports. She worked on the space shuttle and satellites. She worked on a lunar lander, a spacecraft to land on the moon. Her math helped people visit the moon and get home safely. The little girl who loved to count helped change the world.

"I went to work every day for 33 years happy," she said. "I loved going to work every single day."

Katherine retired in 1986. She traveled and spent time with her family and friends. She liked talking to students. She told them to learn more about math and science. She told them to study, work hard, and never give up on their dreams.

She said, "We will always have STEM with us. Some things will drop out of the public eye and will go away, but there will always be science, engineering, and technology. And there will always, always be mathematics. Everything is **physics** and math."

TIME LINE

1918: Katherine Coleman is born on August 26.

1932: Katherine graduates from high school.

1937: Katherine graduates college at age 18. She earns degrees in mathematics and French.

1939: Katherine marries James Goble.

1940: Katherine attends West Virginia University.

1953: Katherine begins work at NACA.

1956: James Goble dies.

1958: NACA becomes NASA.

1959: Katherine marries James Johnson. She becomes Katherine Johnson.

1962: Katherine checks the numbers for the Friendship 7 mission. On February 20, John H. Glenn, Jr., is the first American to orbit Earth.

1986: Katherine retires from NASA.

2015: Katherine gets the Presidential Medal of Freedom. This is one of the greatest honors in America. President Barack Obama calls Johnson, "a pioneer in American space history."

2016: The book and film *Hidden Figures* come out. They follow African-American women who worked at NASA. Katherine is featured with Mary Jackson and Dorothy Vaughan.

2018: Katherine turns 100 years old.

GLOSSARY

analyzed (AN-uh-lized): carefully studied

geometry (jee-AH-muh-tree): a branch of math that deals with lines, angles, and shapes

integration (in-ti-GRAY-shuhn): the act of combining into a whole; to end the separation of people by race

orbit (OR-bit): to go in a curved path around a moon, planet, or other body in space

physics (FIZ-iks): the study of matter and energy

research (REE-surch): related to the careful study of a subject to learn new facts

segregation (seg-ri-GAY-shuhn): the act of separating people or things from the main group

servant (SUR-vuhnt): a person who does housework for other people

INDEX

TEXT-DEPENDENT QUESTIONS

1. Why did Katherine move to go to school?

2. How did people solve math problems before machine computers were invented?

3. How were African-American women treated at NACA?

4. What happened when Katherine started going to meetings?

5. Why did John Glenn want Katherine's help?

EXTENSION ACTIVITY

Katherine said that everything is related to physics, a branch of science, and math. How much science and math can you find in your life? How did they help make the clothes you wear? The foods you eat? The technology you use? Your house and your school? Pick one object and research it. How was it made? Would you have that thing without science and math? Share what you learned about the object.

ABOUT THE AUTHOR

M.M. Eboch also writes books as Chris Eboch. History is one of her favorite subjects. Her book *The Eyes of Pharaoh* is a mystery in ancient Egypt. *The Well of Sacrifice* is a Mayan adventure. She lives in New Mexico with her husband and their two ferrets.

ABOUT THE ILLUSTRATOR

Elena Bia was born in a little town in northern Italy, near the Alps. In her free time, she puts her heart into personal comics. She loves walking on the beach and walking through the woods. For her, flowers are the most beautiful form of life.

www.rourkeeducationalmedia.com

Quote sources: Alexa C. Kurzius, "Hidden Hero," Super Science, Scholastic, (February 2017): https://superscience. scholastic.com/issues/2016-17/020117/hidden-hero.html#800L ;"Katherine Johnson: A Lifetime of STEM", NASA, November 6, 2015, https://www.nasa.gov/audience/foreducators/a-lifetime-of-stem.html ;Michael Mink, "Katherine Johnson Did The Math For NASA When It Counted Most", Investor's Business Daily, December 29, 2016, https://www. investors.com/news/management/leaders-and-success/katherine-johnson-did-the-math-for-nasa-when-it-counted-most/

PHOTO CREDITS: Page 20: ©NASA

Edited by: Kim Thompson
Cover and interior design by: Rhea Magaro-Wallace

Library of Congress PCN Data

Katherine Johnson / M.M. Eboch
(Women in Science and Technology)
ISBN 978-1-73161-427-8 (hard cover)
ISBN 978-1-73161-222-9 (soft cover)
ISBN 978-1-73161-532-9 (e-Book)
ISBN 978-1-73161-637-1 (ePub)
Library of Congress Control Number: 2019932132

Rourke Educational Media
Printed in the United States of America,
North Mankato, Minnesota